SpringerBriefs in Business

For further volumes:
http://www.springer.com/series/8860

Ralf Bebenroth

The Great East Japan Earthquake and Its Impact on German Firms

An Empirical Study

Springer

Ralf Bebenroth
Research Institute for Economics
and Business Administration
Kobe University
Kobe, Japan

Asia Africa Institute on sabbatical leave
Hamburg University
Hamburg, Germany

ISSN 2191-5482 ISSN 2191-5490 (electronic)
ISBN 978-4-431-54450-0 ISBN 978-4-431-54451-7 (eBook)
DOI 10.1007/978-4-431-54451-7
Springer Tokyo Heidelberg New York Dordrecht London

Library of Congress Control Number: 2013943703

Printed on acid-free paper

Springer is part of Springer Science+Business Media (www.springer.com)

Preface

This study aims to investigate the impact of the Great East Japan Earthquake on Kanto based foreign firms. More specifically, this research deals with German firms in the aftermath of the Earthquake on March 11th 2011, including the Tsunami and the meltdown of the Fukushima nuclear plant (in the following called: disaster).

As presented in this booklet, there is quite an amount of literature available about impact on people in the Tohoku region and also on Japanese firms. However, there is a scarcity of academic research about Japan based foreign firms and how the shock of the disaster impacted them.

This booklet attempts to shed some light on this topic through a unique questionnaire-based study and follow-up interviews with top executives at 14 different Kanto based German subsidiaries. Some of the questions to be answered are: To which degree were German firms impacted by the disaster? To what extent did they close temporarily or relocate to other parts of Japan? How did expatriated people and other non-Japanese employees behave in this situation? What is the situation for German subsidiaries nowadays?

Hamburg, Germany, July 2013 Ralf Bebenroth

Acknowledgements

The empirical questionnaire study was conducted on March 2012, one year after the Great East Japan Earthquake. After the return of the questionnaires, the author conducted follow up interviews with top managers of 14 different Kanto based German firms. This was done on June 2012. This booklet contains three chapters which were delivered by invitation of the German Consulate in Osaka at "Wirtschaftskreis" (German scientific roundtable) on November 29th 2012. The three chapters will appear in a somewhat similar fashion in various journals, but in other languages.

The first chapter is about the fundamental situation of the impact on German firms. It is enriched by qualitative information gained at the interviews and can be seen as a more rigorous, developed version of the 6 pages published at Controller Magazine in German with the title: Auswirkungen der Dreifachkatastrophe auf deutsche Unternehmen in Japan, 2012, September, pp. 76–81.

The second chapter deals with "change" and divides subsidiaries who stayed at the Kanto offices and operated "as normal" and others who made some changes in terms of either a temporary closing or even of an office relocation out of the Kanto area respectively. This chapter is in preparation to get published in Japanese language with somewhat similar arguments at the Japanese Business Conference Journal. The second chapter has been improved by thoughtful comments from Dr. Werner Pascha, Prof. at East Asian Institute at University of Duisburg-Essen and was delivered at the same place on April 26th 2013. Also, the second part was improved by the comments of Prof. Dr. Tomoki Sekiguchi when it was delivered at the European Asian Management Conference at Singapore in November 2012. Furthermore, the second part was delivered by invitation of Stockholm University, Sweden in June 2013.

The third chapter sheds light on subsidiary dependency on expatriates and is based, to some extent, on what is forthcoming at Kokumin Keizai Zashi. The idea of this third chapter was first written in Japanese but does only contain few interview statements. It will appear with the title: gaishikei kigyo he no higashi nihon dai shinsai chokugo: chuzaiyakuin he no izon to honsha no kakujitsu ni chakumoku shite (Great East Japan Earthquake and Impact on Foreign Affiliated Firms: Viewpoint of Expatriate Dependency and Headquarter Trouble).

All three of the chapters will be, to some extent, published somewhere else. However, the content of all three chapters in this booklet is substantially further developed compared to the other outlets. Furthermore, all three chapters are to be published in languages other than English; that is, in German and Japanese respectively. Therefore, this booklet hopefully finds its market.

The author would like to thank several people. First, my appreciation goes to the German Chamber of Commerce at Tokyo, which made the empirical study possible through their support in the questionnaire distribution. I was aware that this topic was still quite politically sensitive, and there might have been many good reasons not so support this kind of a study. Without the Chamber's support in connecting me to top executives at German firms, I would not have obtained this respectable response rate of 35 %. Thanks in part to their kind assistance, 86 firms responded to my survey and this led to a representative return.

I also would like to thank Juno Kawakami san from Springer for inspiring and inviting me to undertake this kind of venture to publish a booklet at SpringerBriefs series when I met her at Nihon Daigaku in September 2012. Also I would like to thank my student at Doshisha Business School, Angela Johanna Troianos, for a careful proofread of this script. Furthermore, thanks to my research assistant Kohei Kijima, PhD candidate at Business faculty, on whom I practiced my Japanese, like always, reading to him for many hours Japanese newspaper articles about "Higashi Nihon Dai Shinsai." The time talking to Kijima kun is always fruitful and appreciated.

Finally, it is to mention that this research received financial support by the Japanese government, Grant-in-Aid for scientific research No. 25380504 (Great East Japan Earthquake and Autonomy of Japan Based Foreign Affiliated Firms).

Even though the booklet is a rather small one, it took more of my time and energy to finish it than I anticipated before I started this project.

Contents

Chapter 1
Disaster Impact

Abstract This empirical study investigates the Great East Japan Earthquake of March 11th, 2011 and its impact on German firms located in the Kanto area. It becomes evident that German firms were negatively impacted immediately after the disaster. More than half of all Japan based German firms either closed their offices or relocated to other parts of Japan. In addition, many of the firms reported a high fluctuation in the number of their classic expatriates, short-term visitors, and locally hired non-Japanese. However, 1 year after the disaster, Japan based German subsidiaries have largely recovered. All of the participants stated that they now operate "as normal." Also, they no longer find it difficult to attract people to Japan.

1.1 Introduction

There is much information available about the living conditions of people in the affected Tohoku area, about the temporary shelters where more than 160,000 Japanese citizens still remain 18 months after the disaster. There is also ample information in the Japanese media about the economic impact of the disaster (Petrova 2012, unpublished). The Ministry of Agriculture reports on the environmental damages especially in three regions such as Iwate, Miyagi and Fukushima (Ichise 2011). In addition, research exists on the overall impact for bigger, listed Japanese firms (Tatano et al. 2012, unpublished) and their supply chain problems (Matsuo 2012). There is, however, no research available about foreign-affiliated firms in Japan and how they were impacted by the disaster.

This study will attempt to fill that void. This booklet focuses on the Tohoku disaster from the perspective of Japan-based German subsidiaries. Specifically, it will be explored how German subsidiaries were economically impacted by the disaster. Furthermore, the impact on non-Japanese at the German subsidiaries will be investigated. For this, non-Japanese employees at German subsidiaries will be divided into three groups and the disaster impact on them investigated.

R. Bebenroth, *The Great East Japan Earthquake and Its Impact on German Firms: An Empirical Study*, SpringerBriefs in Business, DOI 10.1007/978-4-431-54451-7_1,

1.2 Firm Level Impact

Available data regarding the impact of the disaster on foreign-affiliated firms is based on case studies obtained mainly from newspapers. IKEA is one example. Their headquarters were originally located at Funabashi (Chiba prefecture), which is close to Tokyo (IKEA also has other Kanto-based store locations and some in the Kansai area). According to a representative of IKEA, the firm decided on March 16th, five days after the disaster, to relocate their employees to Kobe. On average 80–100 employees were temporarily moved to Kobe. Reasons for this decision included a shortage of electricity in the Funabashi headquarters, fear of radiation, and concern about aftershocks (Sato and Haruyama 2011).

The Swedish fashion maker H&M temporarily closed nine of their stores in the Kanto area. Also, their flagship store at Shibuya was temporarily closed and its staff was moved to Osaka. The company explained that they did this because aftershocks distracted employees from their work. The main office was then shifted to a hotel conference room prepared for the firm in Osaka (Ueda and Takeuchi 2011). This make-shift workspace was staffed with 15 H&M employees.

The French tire firm Michelin moved their employees out of the Kanto area, primarily to Kansai. All of their French-expatriated headquarter managers were sent home to France. The same occurred in Philip Morris, the tobacco maker, who also moved their non-Japanese staff temporarily to the Kansai area (Sato and Haruyama 2011).

Also on the level of foreign citizens in Japan, there is some anecdotal evidence available e.g. about Chinese citizens, which make the biggest portion of foreigners in Japan. Their move out of Japan was visible at many locations, especially at the biggest Chinatown in Japan, at Yokohama, geographically close to Tokyo. Before the disaster, this tourist spot employed about 2,500 Chinese citizens. In the aftermath of it, it is reported that around 300 citizens immediately evacuated from Japan because of fear for radiation (Ogura 2011). There is also some evidence available that at Berlitz, one of the biggest language schools in Japan, 30–40 % of their foreign teachers immediately left home after the disaster. However, 5 weeks later almost 90 % of their mostly Anglo Sachs non-Japanese teaching staff seemed to be back at Japan again. In contrast, only very few Chinese came back to Japan. Reasons for this is that mainly of them are students or taking intern. It is reported that their parents hold them back to China. Also one of the biggest fast food chains, Yoshinoya, faced hardship because of their (Chinese employee's) leave. Yoshinoya, which employed around 800 foreign part timers, most of them Chinese, realized that in the aftermath of the disaster a quarter of them evacuated from Japan, back to china (Ogura 2011).

The next paragraph will discuss two questions naturally arise regarding the scope and benefits of this study. First, why focus only on German firms? And second, how can exist information be used to provide a deeper understanding of how these firms were affected by the Tohoku incident?

1.2.1 Why Focus Solely on German Firms?

Japanese newspapers reported broadly, though not necessarily deeply, about German firms during and after the disaster. For example, it was reported that BASF Tokyo changed the location of its headquarters shortly after the earthquake (to Nagoya and Osaka) (Nihon Keizai Shinbun 2011), and that Volkswagen Group Japan sent 9 out of 12 expatriates home to Germany (Yomiuri Shinbun 2011). It is also known from the Japanese news sources that SAP sent its German employees home and relocated its office to Osaka in the days following the tragedy (Ogura et al. 2011), but then returned operations back to Tokyo as early as March 24th (Yomiuri Shinbun 2011). Despite an abundance of facts provided by the Japanese media, it comprises at best only anecdotal evidence consisting of single, ununified cases. The author, therefore, decided to undertake an empirical study of these cases in order to provide a comprehensive and integrated analysis of the overall impact on German firms.

Germany is the second-largest economy in Japan, exceeded only by the United States (Toyokeizai 2012). The resulting magnitude of economic impact incurred by German firms merits this research. Additional reasons for focusing solely on German firms are threefold.

First, the author is a German citizen, so access to German firms was easier than other foreign firms. German language ability proved advantageous in post hoc interviews, during which participants preferred to answer highly sensitive questions in their own mother-tongue. In fact, it greatly reduced the interviewees' anxiety to be able to speak in their native language about highly personal experiences regarding the disaster, their firms, and their families. In addition, the author had established communication with some of the research participants before the disaster occurred. These pre-existing relationships were helpful in encouraging participants to talk about what they regard as a controversial and political issue.

Second, each country reacted differently in the aftermath of the earthquake. Therefore, in order to minimize bias in this research, it was necessary to focus this survey on a single country. In the days, following the earthquake, the French government ordered a complete move of all of their citizens back to France immediately following the incident whereas the British government adopted a very different approach. The British embassy, instead, remained in Tokyo and established a hotline to provide up-to-date information for any non-Japanese citizens who requested it. Furthermore, British citizens were encouraged not to react hastily regarding evacuation of the area. In light of these varying policies regarding evacuation, it was preferable to focus on firms from a single country, such as Germany.

The third reason is the invaluable support of the German Chamber of Commerce in Tokyo. In spite of the political sensitivity of this topic, the Chamber was supportive of this research. It sent a cover sheet (prepared by the author) as well as an invitation to all 244 of its member firms to participate to an internet-based questionnaire. This was helpful not only in terms of time and cost savings, but also because the Chamber used its own network to send messages directly to top executives of all

Kanto-based German firms. As a result, many respondents left their e-mail addresses for a follow-up interview.

Had the author attempted to contact these firms himself, without the Chamber's assistance, responses would have probably come from lower-level employees. It's also probable that those employees might not have even been in Japan at the time of the disaster. And even if they had been, employees from lower levels of the organizational hierarchy are unlikely to understand the whole situation, especially headquarters-related issues. However, thanks to the efforts of the Chamber, follow-up interviews were conducted with top managers of 14 different Kanto based German firms, providing rare insight into challenges German executives faced following the disaster.

In summary, the three reasons listed above explain why only German firms were investigated in this study.

1.2.2 Questionnaire Based Research

This research is questionnaire-based with a range of follow-up interviews. In April 2012 (1 year after the disaster) the electronic questionnaire was prepared by the author in English as well as in Japanese (see Appendix). The targeted firms could choose their preferred language. Out of 87 responses, 86 firms had their main business located in the Kanto area. The remaining firm was not based in the Kanto area; rather, it's headquarter was located in Kobe. Therefore, it was excluded from the sample.

The sample is geographically well positioned because Kanto-based firms felt the impact of the disaster much more than firms from other regions, e.g. from Kansai. The following questions were raised:

- How strong was the impact of the disaster on German subsidiaries?
- How autonomous was the subsidiary against the headquarters?
- How was the impact on non-Japanese employees at the German subsidiaries?

Employees were divided into three categories; classic expatriates, short-term visitors, and locally-hired employees. Classic expatriates are defined as employees who originally worked in the German headquarters but were then dispatched to the Japan-based subsidiary for a time between 2 and 5 years. Short-term visitors resemble managers or technical engineers; however, they were dispatched to the Japan-based subsidiary for a predetermined, temporary time period. These short-term visitors may come often to the Japanese subsidiary but stay there only for a short period of time (usually a period of two to five days). The third category, locally-hired non-Japanese, are locally sourced employees. In contrast to German expatriates, they belong juristically to subsidiaries. This means they are not only hired but also contracted to the local offices and not to German headquarters, resulting in some disadvantages for them, like no or very limited fringe benefits in terms of housing allowance etc. For the purpose of this study, locally-hired non-Japanese are put together to one group, even if some non-Japanese employees (in fact, may be non-German) who came to Japan for other purposes and were eventually hired by the Japan-based German subsidiary.

It should be noted that this category of employee represents a valuable resource for the Japan-based subsidiary because they save the firm the transaction costs associated with dispatching expatriates from Germany. They are also likely to be more knowledgeable about Japanese corporate culture, language and the overall business behavior. The following questions were asked regarding the three categories of employees:

- How serious was the sudden departure of non-Japanese employees from the subsidiaries?
- To what extent did non-Japanese evacuate from Japan even if the subsidiary management wanted them to stay?
- To what extent became sudden departures a problem for the Japanese subsidiaries?
- Do subsidiaries still have difficulties to attract non-Japanese employees (1 year after the disaster)?
- To what extent do foreign firms rely on non-Japanese employees?

1.3 Economic Impact and Move

1.3.1 Temporary Closing or Relocation

This study aims to investigate the situation in which German firms found themselves right after the disaster. Even if occurrences of earthquakes should not surprise in a country like Japan, many firms and their employees were obviously shocked by the aftershocks and radiation scare that followed the March 11th, 2011 disaster (Ueda and Takeuchi 2011). In the questionnaire, three alternatives were offered on how foreign firms treated the situation:

- The firm continued business as usual, operating "normally" without any change.
- The office was at least temporarily closed and employees were sent home. In this case, employees either took vacation or worked remotely at home.
- The office was relocated to another part of Japan. That means that the office obviously saw bigger problems that led them even to relocation. There are some newspaper reports commenting on this.

The empirical study brought following results. Overall, 47 % of the firms continued their operation "as usual." This means that the rest of the firms, or 53 %, did not proceed as usual and undertook some changes either to temporarily close their office or to even relocate their offices to other parts of Japan. Around 45 % of the participating firms closed their offices at least for a while and sent their employees home. The remaining 8 % relocated their offices to other parts in Japan, either to Nagoya or

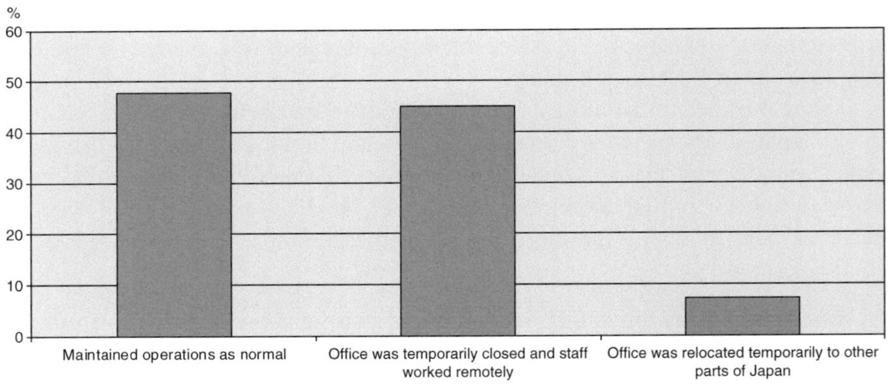

Fig. 1.1 Disaster impact on firms

Osaka. Follow-up interviews revealed that some of these subsidiaries already had pre-existing offices in those areas and simply merged the two offices together.

This situation verifies anecdotal evidence from newspaper reports that, indeed, quite a large number of foreign firms underwent some changes following the disaster. Follow-up interviews revealed that even firms which reportedly "maintained operations as usual" allowed employees the choice to go home at any time. Further evidence that these firms could not operate as usual is provided by logistical disruptions that caused difficulties for employees. For example, trains stopped or ran with special, limited time schedules, making it difficult for staff to commute to the office. Convenience stores faced a shortage of food and of potable water. That strongly suggests that even some firms which intended to "operate as normal" may have had to temporarily close their office or relocate it to another area outside of Kanto (Fig. 1.1).

In addition to temporary closings and relocations, inquiries were made into the disaster impact on individual businesses. Approximately 5 % of participants perceived a "very negative" impact, and 23 % a "negative" impact. Around 46 % of participants reported "no noticeable" impact from the disaster. Surprisingly, 24 % of the firms saw a "rather positive" impact on their business because of the disaster. Follow-up interviews revealed that most of the German firms are engaged in machinery and construction-related businesses; therefore, they predicted that being able to help with the restoration would bring them new business (Fig. 1.2).

Additionally, participants were asked "How do German firms see the impact of the disaster in the long run?" Answers regarding long-term impact were similar to those regarding the short-term. Compared to the short-term impact, there was a 3 % decrease of the firms who saw a "strong negative" impact in the long term, and 12 % which saw a "somewhat negative" impact in the long run. About 58 % saw "no noticeable impact," indicating that the majority of firms do not see any impact in the long run. A high number of firms, 27 %, even saw a positive impact on their business (Fig. 1.3).

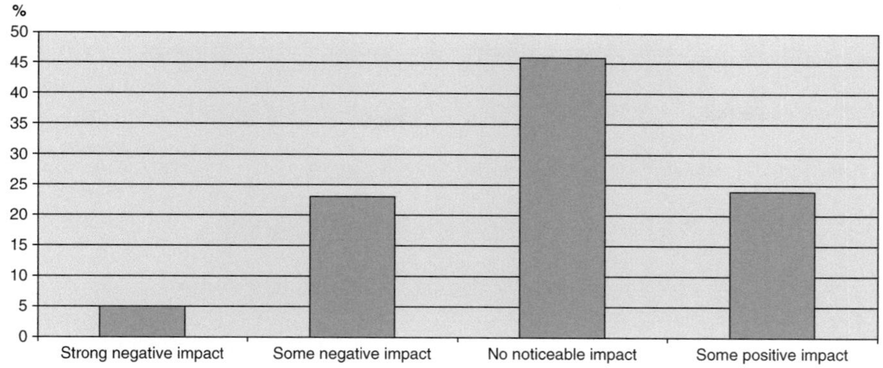

Fig. 1.2 Impact on own business

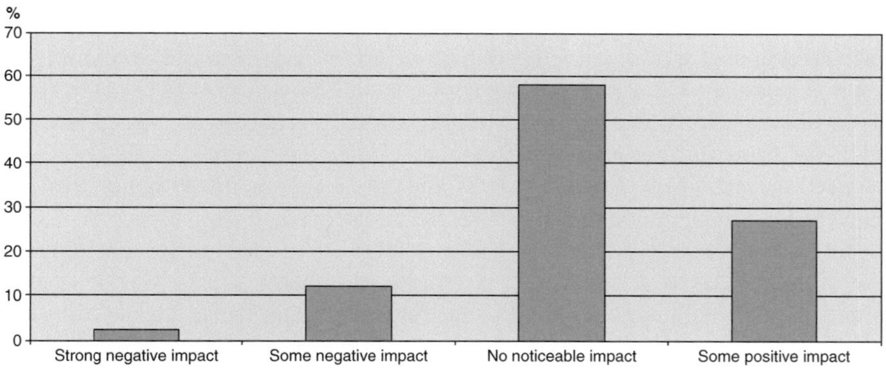

Fig. 1.3 Impact on business in long run

1.3.2 Economic Impact

To obtain more detailed information about the economic impact, several questions were raised about performance items. Performances of the firms were split into several categories such as turnover, market share, profitability, innovation and downsizing activity. Participants were asked how much the disaster influenced their subsidiaries according to these variables. Around 35 % of the firms admitted a decrease of their sales volume. This information can be; however, also commented in a rather positive way. Viewed optimistically, this suggests that 65 % did not see any negative impact on it. Although one firm admitted that it could not reach its 2011 target for sales volume, they recovered much faster than they originally expected.

Overall, less than 10 % of participating firms admitted a loss in their market share. Some of these firms expected to see continued impact until 2012 or later. Profitability goals were reached despite the disaster for almost 70 % of the participants. That is to say, about 30 % of the participants did not reach their forecasted figures. Some of the

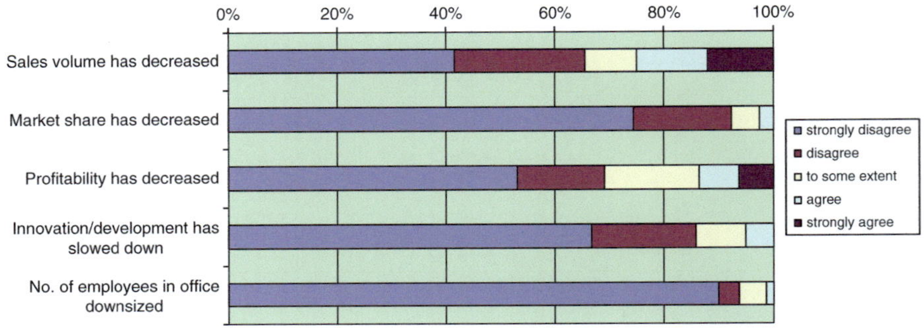

Fig. 1.4 Performance deterioration detailed

respondents stated in the follow-up interviews that profitability decreased as they faced additional costs for implementing better risk management and additional prevention measures. Innovation and development were impacted for 14 % of the firms, where the majority of them say they were impacted "to some extent." Approximately 94 % of firms did not reduce the number of their employees, while only 5 % reduced their numbers "to some extent" and only 1 % (one firm) "agrees" to have reduced their number of employees after the disaster (Fig. 1.4).

Interviews revealed additional problems subsidiaries faced in the aftermath of the disaster, such as a shortage in electricity and energy supply. Saving energy seemed to be a big hurdle for some of the foreign firms in Japan. This is still a hot topic today, not only in the Kanto area but throughout Japan (Hori 2012).

1.3.3 Autonomy and Leaving Japan?

Regarding the decision firms faced about whether to departure or remain in Japan, focus was placed on two different but interrelated questions. First, did German firms leave Japan because of the disaster? Second, how autonomous were Japanese subsidiaries against influences of headquarters?

The first question, on whether or not German subsidiaries continued in Japan, is simple to answer. Around 91 % of all participants answered "fully agree" and 5 % answered with "agree" that they continue their businesses in Japan. The other 4 % answered with "more or less." Participants also disagreed with a 95 % response rate to a relocation to another Asian country. When asked if the office in Japan might have been under pressure to close down, 95 % responded "totally disagree" to this threat.

As the opinions of managers in subsidiaries and headquarters naturally differ, the degree to which the subsidiary remained autonomous was investigated. A high level of regional autonomy suggests that the Japanese subsidiary was given more

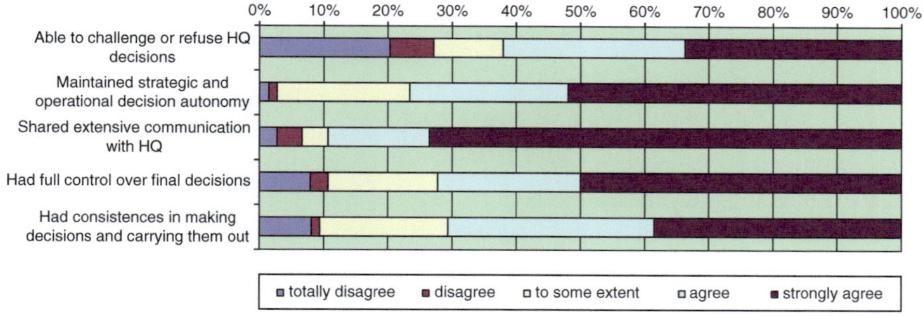

Fig. 1.5 Degree of autonomy

power to decide whether to stay or departure Japan. Participants were asked the following questions:

- Was the subsidiary able to refuse headquarter' decisions?
- Did the subsidiary maintain strategic and operational decision-making autonomy?
- Did the subsidiary share extensive communication with headquarters?
- Did the subsidiary have full control over final decisions?
- Was the subsidiary able to consistently make decisions and to independently carry them out?

The results of this study show that the majority of the firms were able to refuse decisions of the headquarters (62 % of the participants). Only roughly one third (27 %), of the firms were not able to refuse headquarter decisions.[1] The autonomy of strategic and operational decision making was obviously located in Japan. This is verified by 98 % of all respondents, who replied with at least "more or less." Besides autonomy, many firms gave high importance to extensive communication with the headquarters at the aftermath of the disaster. 90 % of respondents replied that they (the Japan-based subsidiary) maintained control over final decisions. This indicates that even if the subsidiaries had high autonomy, they also maintained intensive communication with headquarters. This item was answered with 74 % agreement, of that with 58 %, "fully agree" and a 16 % "agree" respectively. Similar results were seen for consistencies in making decisions and in carrying them out (Fig. 1.5).

In terms of subsidiary relationships with headquarters, a specific questionnaire item was applied. Firms were asked how much trouble top management was thought to have with the subsidiary. The result is that 61 % of the subsidiaries indicated they have a good relationship with their headquarters (they "disagreed" or "totally disagreed" to this item of having any troubled relationship). On the other hand, 28 % faced trouble with their headquarters and 11 % answered the question with "to some extent."

[1] The pattern of English responses; however, was different of the questionnaires filled out in Japanese language. Japanese respondents showed a much lower autonomy and could seldom refuse headquarter decisions.

1.4 Situation of Non-Japanese Employees After the Disaster

Managers were asked about the behavior of their non-Japanese employees in the aftermath of the disaster. Sudden departures of expatriates were frequently based on reasons other than the personal desire of the individuals to return home. Therefore, these extenuating reasons need to be mentioned.

The governments of individual countries exhibited implemented various policies regarding evacuation of their citizens from Japan. As stated in the introduction, firms were impacted by their governments' policies. The British Embassy did not require immediate evacuation of British citizens whereas the French Embassy did. Often, families of non-Japanese employees requested them to return to their home countries. Also, some headquarters requested their employees to come back to Germany. In short, even if some employees actually wanted to remain in Japan, they could not do so and were forced to departure the country. Let us divide these employees now into three groups: classic expatriates, short-term visitors, and locally-hired non-Japanese.

1.4.1 Headquarter Sent Expatriates

The following questions focused on what happened to classic expatriates[2] in the aftermath of the disaster (Fig. 1.6):

- To what extent did expatriates departure their subsidiaries in Japan?
- How difficult was it for subsidiaries to keep expatriates in the Japan-based subsidiary?
- To what degree did subsidiaries face problems because of their sudden departure?
- How difficult do subsidiaries still find it (1 year after the disaster) to get expatriates back to Japan?
- How strongly do subsidiaries rely on expatriates?

Empirical evidence shows that 54 % of the respondents were faced with a sudden departure of expatriates. That means more than half of all the firms replied "agree" to this question, 40 % even strongly and the others at least to some extent.

One firm stated that only two expatriates returned to Germany but the others remained in Japan. Another firm allowed their 120 expatriates who remained in Japan to work "remotely" from home.

The follow up interviews made clear that several German expatriates did not necessarily return to German headquarters. Rather, in some cases, the headquarters dispatched them to other Asian countries. Some expatriates were placed in offices in Singapore, while others left for Hong Kong. The place of stay largely depended what offices the firms held in Asia at the time of the disaster.

A second item also received a high amount of agreement among respondents. More than half of the sample firms confirmed problems in retaining expatriates at

[2] Expatriates are referred to as headquarter sent employees for a time frame of 2–5 years. They are contracted directly to the headquarters and receive high fringe benefits.

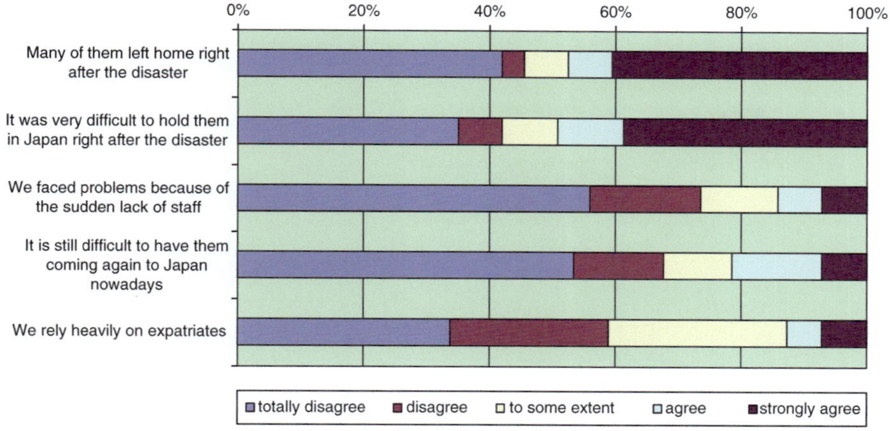

Fig. 1.6 Expatriate behavior

their Japanese subsidiaries. This implies that even if executives located within the subsidiaries wanted to keep expatriates in Japan, they could not do so.

However, a third item reveals that the sudden departure of expatriates did not seem to hurt the subsidiaries. Around 74 % of the participants say that there were no problems caused by the sudden decrease in the expatriate employee pool. To state this differently, only 12 % faced at least some problems and 14 % reported definite problems (answered "agree" and "fully agree"). In the fourth item, it was asked whether it is still today difficult to have expatriates come back to the subsidiaries. Around 32 % agreed to this question with an answer of "to some extent." One firm responded: "[it is] difficult to draw [a] new person to Japan but impossible to get previous expats [back] to Japan."

Therefore, it can also be seen that almost 70 % of German subsidiaries in Japan do not have any difficulties nowadays getting expatriates back to Japan.

Next, the survey examined the degree to which subsidiaries actually relied on their expatriates. The results show that 59 % of the subsidiaries do not feel they rely on expatriates. Even out of the 41 % which rely on expatriates, an impressively high number (29 %) declared that they are only "more or less important" to them.

1.4.2 Short-Term Visitors

Short-term visitors are people who visit the subsidiaries for only a very short time period.[3] As was done for expatriates, the survey also inquired about the degree to

[3] This is often only a few days. Short-term visitors are normally sent from the headquarters or from regional headquarters.

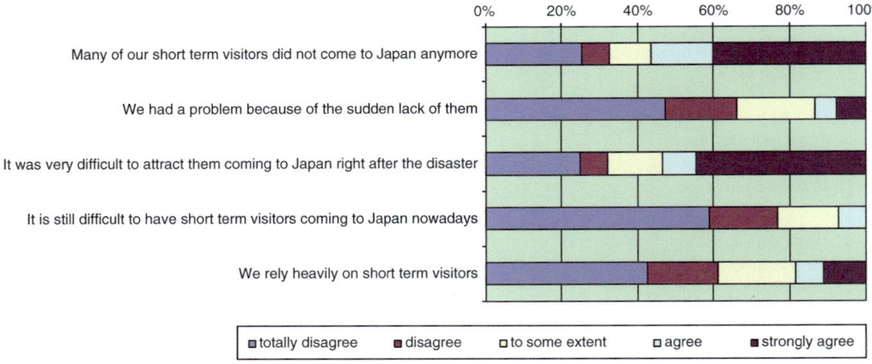

Fig. 1.7 Short-term visitors

which this group was impacted by the disaster. Particular interest was given to the shortage of short-term visitors in the days after the disaster, and whether this caused any problems for subsidiaries. The following questions were raised (Fig. 1.7):

- To what extent did the disaster impact short-term visitors, causing them not to come to Japan?
- To what degree did offices in Japan face problems because of a lack of short-term visitors?
- Was it difficult for the Japanese subsidiary to attract short-term visitors to Japan immediately after the disaster?
- How difficult does it continue to be (1 year after the disaster) to get short-term visitors to Japan?
- How strongly do subsidiaries rely on short term visitors?

As can be seen in the graph above, many of the short-term visitors did not come anymore to the Japanese subsidiary. Around 56 % of the participants supported this statement, 40 % even strongly agreed. In addition, 11 % answered this question with "to some extent." It should be mentioned that official sanctions from the Ministry may have impacted this particular group to a large extent. Even if some of the managers wanted to visit the subsidiary, they might have not been able to go because of official warnings to avoid entering Japan. Japanese offices also state that this type of short-term visitors was also difficult to attract to come again to the subsidiary. Even 54 % "strongly agreed" and another 11 % "agreed" on the item, that it was difficult to attract short-term visitors to come at the aftermath of the disaster again to Japan.

However, similar to the expatriates, participants state that their offices did not face any serious problems because of the lack of them. There is some explanation for this. It might not hurt the Japanese subsidiaries having a lack of them for just some weeks or even month. Only a quarter of the participants faced some problems because of a shortage of short-term visitors, with 11 % of them which answered this item with "more or less."

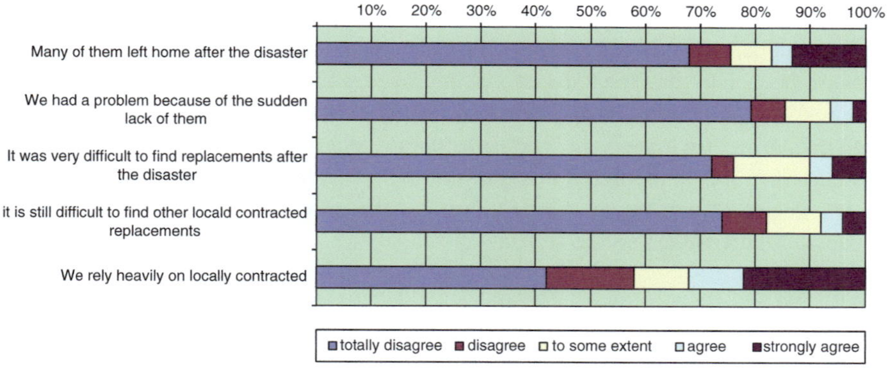

Fig. 1.8 Locally hired non-Japanese

Today it does not seem to be difficult to draw short-term visitors to the Japanese subsidiaries. Around 77 % agree to this, and only 23 % mentioned having at least some problems to attracting short-term visitors back to Japan. Also the item on reliance is similar to the one for expatriates. Only 38 % of the participants admit that they would rely on them, 20 % out of them with only a "more or less."

1.4.3 Locally Hired Non-Japanese

This group of employees in German subsidiaries is not always, but often, composed of German citizens. Some of them could have come to Japan for study/research purposes, or on other programs. Besides a lower income (compared to expatriates), it is arguable that many of them could have left Japan after the disaster, like many expatriates did. If they had done so, they would have faced a variety of unique challenges. One possible problem for these locally-hired employees would have been that they did not have a headquarters to go back to. Because they were hired locally, there was no excuse for them to say the headquarters asked them to come back home. However, it can be assumed that also employees in this category did departure Japan. Participating firms were therefore asked the following questions about locally hired non-Japanese employees (Fig. 1.8):

- To what extent did locally hired non-Japanese managers depart the subsidiaries?
- To what degree did subsidiaries face problems because of their sudden absence?
- Was it difficult for the subsidiaries to find replacements?
- Currently (1 year after the disaster), how difficult is it to get locally hired non-Japanese back to Japan?
- How strongly do subsidiaries rely on locally hired non-Japanese?

Interestingly, some of the locally hired non-Japanese employees also left subsidiaries. At least 25 % of respondents agreed to this item. Within this group, 8 % replied "to some extent." However, there was very little impact regarding problems caused by their departure. Merely 24 % of the participants agreed that it was difficult to replace them. The situation today seems improved, and 82 % of the participants say it is again easy to employ locally hired and locally contracted employees. Notably, a high degree, approximately 42 %, of subsidiaries rely on locally hired Non-Japanese. Among them, 32 % answered this question with "agree" or "fully agree." This is a higher degree of agreement and demonstrates the importance of locally hired non-Japanese relative to expatriated hires.

1.5 Results and Conclusion

The results of this study shows a rather positive picture about the impact of the disaster on German firms. The empirical study indicates that German firms were negatively affected immediately after the disaster. More than half of German firms either closed their offices or relocated to other parts of Japan. However, 1 year after the disaster, it becomes clear that the subsidiaries recovered. All of the participants declared that they now "operate as normal."

Also, although some firms encountered negative effects, a quarter of the subsidiaries argued they have experienced a positive impact on their business because of the disaster.

The biggest group of respondents, half of all the participants, stated that they did not assume any kind of impact on their business due to the disaster. Some firms answered separately that they saw both a positive as well as a negative impact, which ultimately balanced out for them. There is no sign that the disaster forced headquarters to relocate German subsidiaries to other Asian countries. Additionally, subsidiaries in Japan did not close their offices.

German subsidiaries enjoy, on average, a relatively high degree of autonomy. This could be the consequence of long-standing relationships in Japan and long-term agreements with other Japanese firms. In more than half of the cases, expatriates suddenly left the subsidiary. In follow-up interviews it became evident that this decision was often influenced by factors other than the individual's desire to departure Japan. In many cases, the German headquarter called their expatriates back to Germany, and also the Ministry of Foreign Affairs was actively involved in getting expatriates back to Germany.

Japan-based German subsidiaries clearly had difficulties retaining expatriates. Short-term visitors frequently fled to Germany and did not return back Japan for some time after the disaster. However, only very few German subsidiaries reported that they faced economic problems because of that situation. Some locally hired non-Japanese left Japan, albeit to a much lower extent than that of expatriates or short-term visitors. Today, this situation has been normalized. Subsidiaries nowadays find it easier to recruit/attract non-Japanese employees to their subsidiaries.

One might argue that a possible bias exists in this research, because only those German firms which remained in Japan after the disaster were invited to participate in this survey. This would lower the value of the study (because firms replying to the questions would be then more positively exposed than those which closed operations in Japan). However, it should be noted that the number of member firms at the German Chamber of Commerce has not changed more than the normal fluctuation. This indicates that there was no higher-than-normal fluctuation of firms who left Japan. Therefore, this questionnaire study gives a reliable picture of the situation in which German subsidiaries found themselves right after the disaster. Furthermore, it can be shown that normalization of German firms in Japan has taken place again already 1 year after the disaster.

The following two chapters of this booklet contain two separate studies which are based on the descriptive information presented. The second chapter is concerned about temporary closings and relocation issues among German subsidiaries in the aftermath of the disaster. The third chapter deals with issues of expatriates and their sudden departures from Japan.

Appendix: Questionnaire (English Version)

Impact of March 11th disaster on German companies in Japan

One year has passed since the triple disaster of March 11. In cooperation with the German Chamber of Commerce in Tokyo (DIHK), Kobe University is now conducting a scientific survey on the short and long-term impact of the Tohoku earthquake ("disaster") on German companies in Japan.

We are collecting key people's views on the situation of German companies following the disaster. The following questions will be focused on: How much were German companies affected by the disaster? Do they still suffer today or are their operations to normal? Our specific goal is to investigate about human resource aspects. How were different groups affected, such as expatriates from German headquarters, short term visitors or local hired managers?

This study is being carried out by Prof. Dr. Ralf Bebenroth of Kobe University, who holds a tenured position at the Research Institute for Economics and Business Administration.

If you are interested in the survey results, please leave your name and contact at the end of the survey. We reassure you that this survey only serves scientific purposes and will be evaluated anonymously.

The questionnaire will take you not more than 12 minutes to fill out all the questions. Thank you for your cooperation.

Kind regards,
Ralf Bebenroth

(continued)

Appendix (continued)

Questions:

1. How greatly do you sense change at your company because of the triple Tohoku disaster ("disaster") on March 11th?

1.1. What was your company's condition in the weeks after the disaster?

Maintained operations as normal
Office was temporarily closed and staff worked remotely
Office was relocated temporarily to other parts of Japan
Others (please specify):

1.2. How would you describe your current operations?
We operate as normal
Office will be remaining relocated to another Japanese city
Office is relocated to another Asian country
Office is closed
Others (please specify):

1.3. How does the disaster affect your business?
Will have a strong negative impact
Will have a negative impact
No noticeable impact
Will have some positive impact
Will have a strong positive impact
Others (please specify):

1.4. How will the disaster affect your business in the long run?
Will have a strong negative impact
Will have a negative impact
No noticeable impact
Will have some positive impact
Will have a strong positive impact

1.5. How did your company's performance deteriorate immediately after the disaster?
Sales volume has decreased
Market share has decreased
Profitability has decreased
Innovation/development has slowed down
No. of employees in office downsized
Others (please specify):

1.6. Compared to the Tohoku disaster, how serious for your company are the following issues?

(continued)

Appendix (continued)

Strong yen appreciation
High-cost labor in Japan
Trouble solving issues with own headquarter
Others (please specify):

1.7. What would you guess is your headquarters' policy towards your Japan operation?
Headquarter wants to:
Maintain Japan presence as normal
Reduce presence in Japan
Relocate Japan office to another (Asian) country
Close down Japan office
Others (please specify):

1.8. How autonomous would you describe your office in the aftermath of the disaster? Our office management:
Was able to challenge or refuse decisions of the headquarter
Maintained strategic and operational decision autonomy
Shared extensive communication headquarter
Had full control over final decisions
Had consistent procedures in making
Others (please specify):

2. In the following, we would like to ask you some questions about 3 groups of employees: headquarter sent expatriates, short term visitors and about local hired non-Japanese (e.g. German) managers.

2.1. Expatriates

2.1.1. How did expatriates react to the disaster?
Many of them left home right after the disaster
It was very difficult to hold them in Japan right after the disaster
We faced problems because of the sudden lack of staff
It is still difficult to have them coming again to Japan nowadays
We rely heavily on expatriates
Others (please specify):

2.1.2. How important do you consider the following reasons for appointing an expatriate to an executive position?
Establishment and build-up Change in strategy
Crisis management
Development and initial training of local employees
Coordination with mother company Product and sales knowledge Loyalty towards mother company External representation

(continued)

Appendix (continued)

Backflow of information from Japan to headquarter
Japan stay as part of HR development

2.2. Short term visitors

2.2.1. How did short term visitors react to the disaster?
Many of our short term visitors did not come to Japan anymore
We had a problem because of the sudden lack of them
It was very difficult to attract them coming to Japan right after the disaster
It is still difficult to have short term visitors coming to Japan nowadays
We rely heavily on short term visitors
Others (please specify):

2.2.2. How important are short term visitors for the following areas?
Special projects
Strategy
Operative day to day business
Sales Marketing Controlling R&D

2.2.3. Do short term visitors replace or supplement expatriates?
Completely replace
Rather replace
Intermediate
Rather supplement
Only supplement
Don't know

2.3. Local contracted (German, or other non-Japanese) managers

2.3.1. What has changed in regard to local contracted managers?
Many of them left home after the disaster
We had a problem because of the sudden lack of them
It was very difficult to find replacements after the disaster
It is still difficult to find other local contracted replacements
We rely heavily on local contracted managers
Others (please specify):

2.3.2. What are the advantages of having local contracted (e.g.) German managers instead of expatriates from the headquarter?
Lower personnel costs compared to expatriates
Job expertise
Good knowledge of the Japanese market
Convenience I speed of searching and finding personnel

(continued)

Appendix (continued)

3. Basic information about your company:

3.1. What industry does your company operate in?
Chemicals/Pharmaceuticals
Consumer goods (durable/non-durable)
Electrical industry/electronic components
IT telecommunication
Machinery and equipment
Automotive and automotive suppliers
Services (finance, insurance, consulting)
Trading company
Logistics
Others (please specify):

3.2. Company characteristics
What is the nationality of the leading manager in your Japan office?
German Japanese
Others (please specify):
Since when is your ·company active in the Japanese market?
Year:
Where is your office located?
City:
What is your current legal status of your company?
100 % Subsidiary
JV
Representative office

3.3. Employee characteristics
How many employees do you have in your company globally?
How many employees do you have in your office in Japan?
How many employees were in your office in Japan before the disaster?

3.4. Respondent characteristics
What is your nationality:
German, Japanese, Others (please specify):
What is your age:
To complement the survey, we plan to conduct interviews with managers of German companies in Japan. May we contact you for an additional interview (1 hour)?

Yes
No
If you are willing to receive the survey results, please enter your contact data in the box below:

(continued)

Appendix (continued)

Name and Company:
Thank you very much for taking the time to answer this survey.
Ralf Bebenroth,
Research Institute for Economics and Business Administration Kobe University

References

Hori S (2012) Enerugi mikusu nit suite (The challenge of optimal energy mix). Nihon Enerugi Gakkai Shi 91(1):2–9

Ichise Y (2011) Higashi Nihon Daishinsai ni yoru nogyo higai to fukkou no kadai (Agricultural issues and recovery from Great East Japan Earthquake). Norin Kinyu 64(8):42–54

Matsuo H (2012) Supply chain disruptions from the Tohoku Earthquake: what happened to automotive semiconductors and what should be researched? In: Proceedings of the 4th production and operations management world conference 2012, Amsterdam

Nihon Keizai Shinbun (2011) Kansai no chintai offisu (iten juyou) nobiru gaishikei chuushin, tankiriyo he jouken kanwa mo (Increase of temporary demand in rental office space in Kansai area by foreign companies). 6 April, p 3

Ogura S (2011) Higashi nihon dai shinsai: modoranu chugokujin rodousha genpatsu jiko de kokugai ryushutsu (Great East Japan Earthquake: Chinese labor do not want to come back to Japan, departure because of Fukushima incident). Mainichi Shinbun, 25 March, p 4

Ogura Y, Oda K, Tani Y (2011) Higashi Nihon Daishinsai: gaishikei, Tokyo dasshutsu (genpatsu jiko no eikyo shinpai) [East Japan Earthquake: foreign companies out of Tokyo (worries about effect of atomic accident)]. Mainichi Shinbun, March, p 6

Sato A, Haruyama Y (2011) Gaishikei kigyou, Kansai ni sokai (Foreign affiliated firms move to Kansai). Asahi Shinbun, p 8

Toyokeizai Shinbunsha (2012) Gaishikei kigyou soran. Tokyo

Ueda T, Takeuchi A (2011) Higashi Nihon Daishinsai: gaishi, shutokaihi honsha kinou, Osaka ni ichiji iten (East Japan Earthquake: temporary move of foreign companies form the Capital to Osaka). Mainichi Shinbun, 18 March, p 25

Yomiuri Shinbun (2011) Genpatsu jiko ni binkan, gaishikei shain kikoku no ugoki (Sensitive agains atomic accident, transfer home of employees from foreign companies). Yomiuri Shinbun, 7 April

Chapter 2
Subsidiary Autonomy and Expatriate Retreat

Abstract This second chapter examines how two factors, subsidiary autonomy and the retreat of expatriates, influenced Japan-based foreign subsidiaries to either temporarily close their offices or relocate them out of the Tokyo (Kanto) area. It can be shown that subsidiaries with higher degrees of autonomy changed their status significantly more often. This means that the more autonomous the subsidiaries were, the more costly it was for the headquarters in the aftermath of the disaster. In contrast to the transaction cost theory, which argues that better performing subsidiaries have higher levels of autonomy, my findings indicate that a lower level of autonomy would be better to prevent unnecessary closings or relocations in the case of a crisis.

2.1 Introduction

The aim of this study is to investigate the impact of the Tohoku- earthquake of March 11th, 2011 on German firms operating in Kanto area. This research discusses the impact of the disaster based on transaction cost theory. Also, expatriates and their often sudden departure right after the disaster are investigated in light of transaction cost theory. There is much information available about citizens affected around Tohoku area. Japanese media also informs about the economic impact for bigger Japanese firms. Some firms relocated facilities and human resources to other parts of Japan. Also, Japanese firms moved their IT information centers from the Kanto area to the West of Japan. According to Softbank Telecom, which, prior to the disaster, received only one request monthly to set up an IT information center in the West of Japan, requests increased to more than 10 cases a month following the disaster. Two dominant locations after the disaster for IT information centers were Osaka and Fukuoka. Also, NTT West Japan recorded a five- time increase in the number of Kanto firm requests to set up their IT information centers in Osaka (Utsunomiya 2011). However, there is a dearth in research about Japan based foreign firms and

R. Bebenroth, *The Great East Japan Earthquake and Its Impact on German Firms: An Empirical Study*, SpringerBriefs in Business, DOI 10.1007/978-4-431-54451-7_2,

how they managed the shock in the aftermath of the disaster. All that is available about foreign affiliated firms are case study level reports based mainly on anecdotal evidence from newspapers. Japanese newspapers report about German cases, that e.g. BASF Japan headquarter changed its location shortly after the disaster from Tokyo to Nagoya and Osaka (Nihon Keizai Shinbun 2011). Volkswagen Group Japan sent 9 out of 12 expatriated home to Germany (Yomiuri Shinbun 2011). Also SAP sent its employees temporarily to Germany and relocated the office temporarily to Osaka (Ogura et al. 2011), but returned to Tokyo as early as March 24th (Yomiuri Shinbun 2011).

This study investigates a sample of Japan- based German subsidiaries to see which factors influenced German affiliated firms to temporarily close down or relocate their subsidiary offices out of the Kanto area. Two topics are explored: the degree of autonomy at the subsidiary and the retreat of expatriates who often suddenly left home to Germany. Top managers of 86 different firms responded to the questionnaire. A main question was "Did your office temporarily close" or "relocate" to other areas in the aftermath of the disaster? Through missing values, 70 firms responded to the question of change in status after Tohoku disaster. Out of these responses, 33 continued their operations after Tohoku disaster "as usual" and did not undertake any change at all. Out of the remaining 37 firms, 32 firms temporary closed their offices for at least several days or weeks and 5 temporarily relocated their offices to Kansai.

As seen in Chap. 1, a year after the disaster, all of the 86 firms stated that they returned to normal operations. Additional interviews with top managers undertaken by the author provide evidence that a change (either through a temporary closing or a relocation to another geographical area) was costly and, when seen from today's perspective, unnecessary. Many interviewed managers agreed that it should have been prevented. Any kind of change was costly, bounded resources, and often led to a loss of their clients' trust. Also, expatriates who left Japan now sometimes have difficulties reintegrating themselves into the Japan office. Interviews showed that two distinct groups emerge: the ones who stayed in the aftermath and the ones who left. Some of the interviewees even spoke about a situation closely resembling a war between these two groups.

This research will focus on two questions. First, did the level of autonomy influence the decision of a subsidiary to either temporary close it down or to move their facility to other parts of Japan? Second, did the retreat of expatriates influence the decision to either temporarily close it down or to move their facility to other parts of Japan?

The remainder of this second chapter is as follows: after presenting the theory and development of the hypotheses, methodology is discussed. After that, results are presented and a discussion follows. The chapter ends with a conclusion.

2.2 Theoretical Background

Transaction cost theory in international business is frequently applied to broaden the cost considerations, e.g. relates costs to a given price (Hennart 1991). Reducing transaction costs is based on managing intra-MNC relationships (Birkinshaw et al. 2004;

Young and Tavares 2004). Autonomy of a subsidiary shows the degree to which it can make its own decisions independent of its headquarters (McDonald et al. 2008; Young and Tavares 2004). There is evidence that subsidiaries have, on average, a high degree of autonomy when located further away from the headquarters, in another country (Egelhoff et al. 1998). Evidence also suggests that this is associated with an improved performance outcome (Birkinshaw 1996; O'Donnell 2000). Furthermore, the OLI (ownership, location, and internalization) view by Dunning predicts that developing autonomy can improve performance of MNC development by reducing transaction costs (Dunning 2000).

The literature predominantly suggests that increasing autonomy leads to lower transaction costs for intra-MNC activities. This might be due to shorter and simpler chains of command and increased capabilities to explore e.g. new ventures or new markets. This in turn will lead to increased learning, innovation, and access to desirable resources, which will result in lower transaction costs. However, autonomy can also have a detrimental effect on the performance of MNCs if foreign based subsidiaries are granted too much autonomy. McDonald et al. (2008) state that too much autonomy may lead to inappropriate allocation of resources, or to an over-use of the headquarters' resources (Almeida and Phene 2004; Mudambi and Navarra 2004).

In the expatriate literature, too, transaction costs theory is a widely discussed lens. Most of the authors have applied transaction costs theory to global staffing decisions and broadened the cost considerations. Some of the topics are related to the control, relocation and support of the expatriates, the hiring and training of locally hired non-Japanese managers, and the development of local and international markets (Brewster et al. 2005; Richards 2001; Perez and Pla-Barber 2005; Schaffer and Rhee 2005; Tan and Mahoney 2006), Expatriate managers are highly valued to perform strategic functions, provide high quality communication between the subsidiary and parent company, and to manage crises in the interests of the parent company (Bruning et al. 2011). Transaction Cost Theory is used to investigate the extent to which a correlation exists between a temporary closing/relocation and expatriate issues (for example, sudden expatriate retreats, difficulties retaining employees, and problems caused at the subsidiaries because of their departure).

In this research, two areas are investigated in the frame of transaction cost theory. The first is degree of autonomy. Did increased autonomy at a given Japan -based German subsidiary lead to changes of their status? Put another way, were subsidiaries with a higher amount of autonomy more rigorous in making a decision to temporarily close down or to relocate the office? The second is the retreat of expatriates. Did the retreat of expatriates' impact the behavior of the subsidiaries, causing them to either temporarily close or to relocate their office?

2.3 Development of Hypotheses

In Japan, like in any other country, some subsidiaries are more autonomous than others. Especially in the case of geographically distant German subsidiaries, which can be assumed to have a relatively high autonomy.

As discussed at Young and Tavares (2004) higher autonomy at a subsidiary will increase intra-MNC relationships and, therefore, will lead to higher action of the subsidiary top management. In this case, a subsidiary might opt favorably for a temporary closing or for a relocation of a firm outside Kanto area. That is, there is reasonable argument that higher autonomy would have more likely led firms to change their status in the aftermath of the Tohoku disaster. The subsidiaries with a higher autonomy in their decisions would have taken action. This leads to the first hypothesis:

Hypothesis 1
The higher the autonomy in the Japanese subsidiary, the higher the change.

About half of the subsidiaries in the sample complained about the fact that their expatriates left the office suddenly. This fact is supported by follow-up interviews with top managers of 14 different firms based in the Kanto area. Departure of expatriates might have caused problems. There is reason to assume that departure of expatriates has a negative impact on the change of the subsidiary status. This means that the more the expatriates departure the subsidiary, the more likely should be a change at the office. That leads to the second hypothesis:

Hypothesis 2
The higher the difficulties with expatriate retreats, the higher the change.

Older subsidiaries should have more autonomy. This is based on the experience of the managers at the subsidiary. At older firms, managers should have longer average tenure at the firm. Also, in the case of older subsidiaries, the headquarters should be more distant from and trusting of the management at the Japan based firm. The same arguments apply to expatriates. Older firms in Japan should be bigger in size and have more experience. As the headquarters is more traditional, they should send more expatriates to the Japan subsidiaries. There is, therefore, reason that older subsidiaries employ more expatriates. In subsidiaries with a relatively big group of expatriates, we expect there to be more communication between them. That leads expatriates to become an autonomous group contrasting themselves against locals. The stronger network at older and bigger firms has advantages for expatriates because they can help each other in case anyone faces trouble. However, in a situation of a disaster, a relatively autonomous group of expatriates might make a decision without consulting locals. The decision might be taken in a hurry. In all, there is reason to think that expatriates in older firms should departure Japan earlier than at younger firms. The departure of expatriates would force the subsidiary to temporarily close or to relocate. That leads to the following two hypotheses:

Hypothesis 3a
The older the firms with more autonomy, the higher the change.

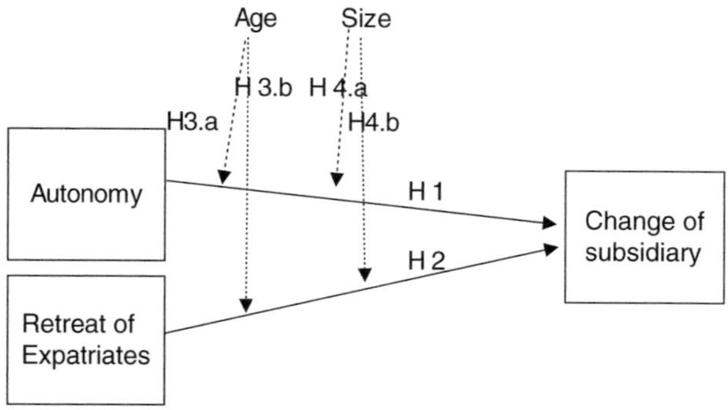

Fig. 2.1 Autonomy and expatriate retreats on subsidiary change

Hypothesis 3b
The older the firms with more expatriates employed, the higher the change.

Similar arguments are developed for the size of the firms. Relatively big subsidiaries should have a higher autonomy. Big subsidiaries have power against the headquarters and a more solid standing to defend their own decisions. They can make decisions on their own. Also here, a similar argument is provided for expatriates and size. Relatively big subsidiaries have more expatriates. When the expatriates suddenly departures the subsidiary, there is a high chance of a change. This leads to the following two hypotheses:

Hypothesis 4a
The bigger the firms with more autonomy, the higher the change.

Hypothesis 4b
The bigger the firms with more expatriates, the higher the change (Fig. 2.1).

2.4 Methods

A questionnaire was developed and administered on March 2012, 1 year after the Tohoku disaster, and sent out to top managers of all German firms in the Kanto area of Japan to share their experiences of the Tohoku disaster and the impact on them. The questionnaire was developed based on exploratory research conducted by the author in Kobe and in Tokyo in February 2012. In all, 244 Kanto based German firms were contacted.

Table 2.1 Results of exploratory factor analysis

	No. of items used in the survey	No. of items used after final iteration	Cronbach's alpha
Autonomy	5	3	0.70
Expatriate retreat	5	3	0.82

The distribution was done by support of the German Chamber of Commerce in Tokyo. An electronic message from the Chamber was sent to top managers of all member firms to ask for their contribution to the project. The pre-tested questionnaire was back translated by two native Japanese speakers and sent out in English and in Japanese. This ensured managers could choose their preferred language to answer the questionnaire. Out of the 244 questionnaires 86 returned, a 35.2 % rate. From these, 70 turned out to be usable for the analyses as there were missing values, especially in the area of expatriates.

Multiple items were used to measure the two independent constructs of the analyses, such as "autonomy" and "retreat of expatriates." Thus, exploratory factor analysis (EFA) of the items was employed to measure both constructs. As suggested by Rai et al. (1996) a method of iteration was employed to derive to a stable factor structure. I eliminated items with loadings of less than 0.5 (Churchill 1979) or with cross-loadings on two or more factors greater than 0.35 (Kim and Mueller 1978). The appendix presents the scales and items. Both constructs were measured with metric five-point Likert scales (strongly disagree = 1, strongly agree = 5).

For the construct autonomy, there are items like: "our subsidiary management was able to challenge or refuse decisions of the headquarters." The first construct resulted in a three item solution having a Chronbach's alpha reliability estimate of 0.70. The second construct "retreat of expatriates" yielded again a three item solution. Questions were based on how seriously the firm was affected by the retreat of expatriates. One of the three items e.g. was: "Many of them left home right after the disaster." This construct resulted in a Chronbach's alpha reliability estimate of 0.82, both exceeding the threshold of Nunnally (1978).

The control variables are age and size. Age is measured in years and shows how many years a subsidiary is already located in Japan. Size is measured as the relative size in log of employees. For this, headquarter employees were divided by the number of employees in the subsidiary to get to a relative number. I then took the logarithm as the differences between firms in sizes in the sample would have been too large otherwise.

For statistical analyses, logit regression analysis was applied having the subsidiary as 0 at no changes and 1 if changes occurred in form of either a temporary closing or a relocation out of the Kanto area. Almost half the firms in the sample, 33 "maintained operation as normal." In contrast, 32 at least temporary closed and 5 moved out of the Kanto area (Table 2.1).

Table 2.2 Correlations

	1	2	3	4	5
1. Change situation	1				
2. Years	0.127	1			
3. Size	0.024	−0.18	1		
4. Expatriates	−0.008	0.10	−0.2	1	
5. Autonomy	0.363	0.29	−0.1	−0.08	1

2.5 Results

I investigated how much autonomy and how much the retreat of expatriates influenced a given subsidiary to either close down the office temporarily or to move it to another part of the country in the aftermath of the Tohoku disaster. Descriptive results show that older subsidiaries are more autonomous (Table 2.2: 0.29). Also, autonomous firms correlate to higher changes (Table 2.2: 0.36). As a result of the logit regression analyses, there is evidence that autonomy influenced changes significantly (H1) but not the retreat of expatriates (H2). Furthermore, high autonomy subsidiaries forced changes especially at bigger firms. Also, surprisingly, it is evident that younger high autonomy firms were significantly related to a higher chance of change of the status in the German subsidiaries in Japan (Table 2.3).

2.6 Discussion

Two constructs, such as subsidiary autonomy and retreat of expatriates, were tested to investigate their impact to a change of German subsidiaries in Japan. About half of the sample opted for a change in status, meaning to temporarily close down and work remotely, or to even relocate the office to other areas outside of the Kanto region.

German subsidiary managers with more autonomy from headquarters clearly showed a higher willingness to undertake changes. This finding could suggest that the subsidiaries with less autonomy were asked from their headquarters to continue as usual in the aftermath of the Tohoku disaster. Also, as stated in the hypotheses, bigger firms with more autonomy changed more often. In contrast to the idea that older firms would change their status more often, there is evidence that it was younger firms who did so. That is to say, the younger the firms, the more often they changed their status to either temporarily close their office or to move to another region. This balances the main finding. The author received similar opinions in interviews from top managers. Often German top managers with a long record in Japan stated that they were the last who wanted to temporarily close down or relocate their activities to other areas, but the headquarters the first.

The retreat of expatriates did not show any impact. The results of this study contradict existing literature which supports the idea that expatriates are given a high influence in moderating relationships between subsidiary and the headquarters

Table 2.3 Logit regression results (Change of subsidiary as dependant variable)

	Model 1		Model 2		Model 3		Model 4		Model 5		Model 6	
	Coefficient	SE	Coefficient	SE	Coefficient	SE	Coefficient	SE	Coefficient	SE	Coefficient	SE
Years	0.01	0.01	–0.00	0.01	0.00	0.01	–0.01	0.01	–0.01	0.01	**1.97**	**1.02+**
Size	0.12	0.44	–0.62	0.60	0.28	0.50	–0.31	0.70	–9.36	5.76	**–19.17**	**9.09***
Expatriates			–0.19	0.27			–0.20	0.31	–0.30	2.15	**–11.89**	**6.53+**
Autonomy					**0.79**	**0.33***	0.78	0.41	–4.50	2.87	–4.70	3.44
Expats*Autonomy							–0.69	2.68	0.20	0.38	**2.79**	**1.47+**
Expats*Size									–0.32	0.54	1.36	1.07
Autonomy*Size									**2.25**	**1.31+**	**3.76**	**1.78***
Expat*Years											–0.10	0.07
Autonomy*Years											**–0.40**	**0.20***
No of observations	50		38		49		38		37		37	
Pseudo R^2	1.1		3		11.7		12.3		24.7		58.2	

+=10 %; *=5 %; **=1 %

* + p<0.1; *p<0.05; **p<0.01; p<0.001

(Bruning et al. 2011). The behavior of expatriates did not influence the subsidiary's decision to temporarily close down the office or to relocate it to another geographic area. The following reasons might explain this. First, expatriates account for a smaller number of employees compared to the full workforce in a Japan-based German subsidiary. Second, because of their low power, the influence of departing expatriates might only very weakly affect the subsidiary. Third, the main job of expatriates is information transmitter to and from the headquarters. Finally, expatriates sent from the headquarters are normally only in Japan for a short while. They are often not the decision makers. Instead, decisions are taken by locally hired Non-Japanese managers with a long track record in Japan. Some of these reasons might explain that the connection between expatriate retreats and a temporary closing or even a move out of the Kanto area is rather weak.

There is evidence from the analyses that a subsidiary with more autonomy made changes or at least convinced the headquarters that a change was necessary. Contrary to the transaction cost theory, which argues that a relatively high autonomy at a subsidiary improves their performance, one might argue that higher autonomy led subsidiaries to make changes which negatively impacted their performance. Subsidiaries in a situation of an external crisis might be better off without too much autonomy. This research provides evidence above McDonald et al. (2008) not only that higher authority leads to worse performance, but also that this is particularly true in case of a crisis.

High autonomy at a subsidiary showed impact for changes in the status. Therefore, autonomy of local management at foreign firms has implications for its success and needs to be studied. Headquarters are advised to consult earlier with their subsidiaries in the case of a crisis. It might be advisable not to give subsidiary management too much authority during a crisis and to have an intensive exchange of information.

This research is not without some shortcomings. First, the questionnaire was only sent out once. Therefore, like all other cross sectional studies, this research has the shortcoming that cause and effect could be reversed. A second shortcoming is that only German firms were asked to participate. There are two advantages with this approach, however. First, through the active support of the German chamber of Commerce which sent a link directly to all its members, the top managers of 86 German subsidiaries could get involved into this research. The electronic distribution ensured that the questionnaire was distributed directly to top managers and filled out by them personally. About 25% of the participants agreed to have additional interviews. After about 14 follow up interviews, the author was assured that top executives (normally the leading person in the subsidiary) answered the questionnaire him /herself. Otherwise, the study might have lacked involvement of the appropriate persons. In particular, questions regarding the subsidiary's autonomy could not be answered by secretary staff or junior management not involved with the actions directly. Another point of discussion is that governments sometimes had quite different policies regarding their countries' firms and their expatriates in the aftermath of the Tohoku disaster. The French government e.g. ordered more or less everyone immediately home to France. Fortunately, as this research deals only with

German firms, the problem of having to address different political influences is resolved. Finally, both constructs such as autonomy and retreat of expatriates were not found in previous research. Even if the two constructs are identified by exploratory factor analyses, the items were defined by the author and might not be similar to other studies of autonomy and expatriate research.

In spite of its shortcomings, this study sheds light on the impact of the Tohoku disaster on foreign firms with theoretical advancement arguing against the mainstream of Transaction Cost Theory application. In this sense, too much autonomy at a subsidiary in a special case of a crisis should be prevented.

2.7 Conclusion

This second chapter dealt with subsidiary autonomy and retreat of expatriates. It investigated how Japan-based foreign subsidiaries were influenced to either temporary close down their offices or to move it out of Tokyo (Kanto) area.

The research shows that higher autonomous subsidiaries changed their status significantly more often—and these firms were the younger ones, not the older, as suggested in the hypothesis. Also, the follow up interviews made clear that top managers now, a year after the disaster, consider a move out of Kansai area or a temporary closing as a costly mistake. Contrary to the transaction cost theory, this suggests that better performing subsidiaries have higher levels of autonomy, I can show that at a time of a disaster a lower level of autonomy would be the more beneficial choice. Managers would not have made these decisions without consulting the headquarters office.

Appendix: Questionnaire Item and Constructs

Item: Change of subsidiary

What was your company's condition in the weeks after the disaster?

- Maintained operations as normal
- Office was temporarily closed and staff worked remotely
- Office was relocated temporarily to other parts of Japan
- Others:

Construct: Autonomy

How autonomous would you describe your subsidiary?

Our subsidiary management:

- Was able to challenge or refuse decisions of the headquarter
- Maintained strategic and operational decision autonomy
- Had full account of final decisions

(continued)

Appendix (continued)

Cronbach's alpha 0.70
Construct: Retreat of expatriates
How serious were your subsidiary affected by the retreat of expatriates?

- Many of them left home right after the disaster
- It was very difficult to hold them in Japan right after the disaster
- We faced problems because of the sudden lack of them

Cronbach's alpha 0.82

References

Almeida P, Phene A (2004) Foreign owned subsidiaries and knowledge creation: the influence of MNC and host country innovation. Strateg Manag J 25:847–864

Birkinshaw J (1996) How multinational subsidiary mandates are gained and lost. J Int Bus Studies 27(3):467–495

Birkinshaw J, Hood N, Young S (2004) Subsidiary entrepreneurship, internal and external competitive forces, and subsidiary performance. Int Bus Rev 14(3):227–248

Brewster C, Sparrow P, Harris H (2005) Towards a new model of globalizing human resource management. Int J Human Resource Manag 16(6):949–970

Bruning NS, Bebenroth R, Pascha W (2011) Valuing Japan-based German expatriate and local manager's functions: do subsidiary age and managerial perspectives matter? Int J Human Resource Manag 22(4):778–806

Churchill AG (1979) A paradigm for developing better measures of marketing constructs. J Mark Res 16(1):64–73

Dunning J (2000) Regions, globalization, and the knowledge economy: the issues stated. In: Dunning J (ed) Regions, globalization, and the knowledge economy. Oxford University Press, Oxford, pp 1–14

Egelhoff W, Gorman L, McCormick S (1998) Using technology as a path to subsidiary development. In: Birkinshaw J, Hood N (eds) Multinational corporate evolution and subsidiary development. Macmillan, Basingstoke

Hennart JF (1991) The transaction cost theory of joint ventures: an empirical study of Japanese subsidiaries in the United States. Manag Sci 37:483–497

Kim JO, Mueller CW (1978) Factor analysis: statistical methods and practical issues. Sage, Beverly Hills

McDonald F, Warhurst S, Allen M (2008) Autonomy, embeddedness, and the performance of foreign owned. Multinational Bus Rev 16(3):73–92

Mudambi R, Navarra P (2004) Is knowledge power? Knowledge flows, subsidiary power and rent-seeking within MNCs. J Int Bus Studies 35(5):385–406

Nihon Keizai Shinbun (2011) Kansai no chintai offisu (iten juyou) nobiru, gaishikei chuushin, tankiriyo he jouken kanwa mo (Increase of temporary demand in rental office space in Kansai area by foreign companies). 6 April, p 3

Nunnally JC (1978) Psychometric theory. McGraw-Hill, New York

O'Donnell S (2000) Managing foreign owned subsidiaries: agents of headquarters or an independent network? Strateg Manag J 21:525–548

Ogura Y, Oda K, Tanita Y (2011) Higashi nihon daishinsai: gaishikei, Tokyo dasshutsu (genpatsu jiko no eikyo shinpai) [East Japan Earthquake: foreign companies out of Tokyo (worries about effect of atomic accident)]. Mainichi Shinbun, 22 March, p 6

Perez J, Pla-Barber J (2005) When are international managers a cost effective solution? The rationale of transaction cost economics applied to staffing decisions in multinational corporations. J Bus Res 58:1320–1329

Rai A, Borah S, Ramaprasad A (1996) Critical success factors for strategic alliances in the information technology industry: an empirical study. Decision Sci 27(1):141–155

Richards M (2001) U.S.multinational staffing practices and implications for subsidiary performance in the U.K and Thailand. Thunderbird Int Bus Rev 43(2):225–242

Schaffer RA, Rhee JH (2005) Consider cost and strategy when choosing between expatriate and host-national managers. J Bus Manag 11(1):59–71

Yomiuri Sinbun (2011) Genpatsu jiko ni binkan, gaishikei shain kikoku no ugoki (Sensitive against atomic accident, transfer home of employees from foreign companies). Yomiuri Shinbun, 7 April

Tan D, Mahoney JT (2006) Why a multinational firm chooses expatriates: integrating resource-based, agency and transaction cost perspectives. J Manag Studies 43(3):457–484

Utsunomiya Yuichi (2011) Higashi Nihon Dai Shinsai: Jouhoukyoten mo nishi nihon he natsu no teiden risuku kaihi (Great East Japan Earthquake: information center move to the west. Prevention of summer shortage energy risk). Mainichi Shinbun, 14 March, p 7

Young S, Tavares A (2004) Centralizafion and autonomy: back to the future. Int Bus Rev 13(3): 215–237

Chapter 3
Subsidiary Dependency on Expatriate Departure and Headquarter Trouble

Abstract This third part deals with expatriate dependency, headquarter trouble, and the impact on German firms caused by the Great East Japan Earthquake. It is evident that firms with a higher dependency on their expatriates experienced fewer employees leaving the subsidiaries. Also, dependency of headquarters on expatriates was positively related to difficulties of retaining those expatriates in Japan. Even today, it is more difficult for firms who rely on expatriates to get them back again to Japan.

Also, research related to troubled headquarter-subsidiary relationships brings two results. First, subsidiaries with higher headquarters trouble faced more difficulties holding on to their expatriates. Second, subsidiaries with a higher headquarter trouble report lower problems in case expatriates left Japan.

3.1 Introduction

This third chapter deals with the Great East Japan Earthquake and impact to expatriates. In this part, subsidiary dependency on expatriates and subsidiary trouble to the headquarters will be investigated. The Social Exchange Theory is used for explanations.

Until now, there has been no scientific evidence available and only some anecdotal evidences by newspaper articles about the disaster impact on expatriates of foreign firms. This research investigates expatriates and how they left the country in the aftermath of the earthquake. It will question why they left; however, it will investigate what the subsidiaries in Japan felt about their sudden departure. Also this study is drawn out of the answers of the questionnaire sample sent by the author to all listed 244 German firms with support in distribution by the German Chamber of Commerce at Tokyo. The questionnaires were sent on March 2012 with a return of 86 complete filled out questionnaires. A month later, on May 2012, follow up

R. Bebenroth, *The Great East Japan Earthquake and Its Impact on German Firms: An Empirical Study*, SpringerBriefs in Business, DOI 10.1007/978-4-431-54451-7_3,

interviews were conducted by the author with top managers of 14 different German firms located in Kanto area.

It is no surprise that the Great East Japan Earthquake damaged foreign affiliated firms in Japan. As seen in the second chapter, half of the sample firms temporarily closed their Kanto based offices or even relocated their businesses to other parts of Japan. Some others just continued their operations as normal. However, even if firms just stayed and "operated as normal," information taken from interviews support that there was some negative economic impact (as discussed in Chap. 1 of this booklet). According to interview comments, these higher costs were associated with an additional risk management and other costs in regard to prevention measures against a similar disaster for the future (Bebenroth 2012).

This third chapter is based on Social Exchange Theory as its theoretical background. After discussing expatriate issues with this theory, hypotheses will be developed and tested with a multi regression analysis. Findings will be presented and a conclusion will be provided.

3.2 Theoretical Background

This chapter applies the Social Exchange Theory in investigating the impact of the Great East Japan Earthquake on German firms. This theory analyses the context where social exchange takes place (Blau 1964). Social Exchange Theory is widely applied and attempts to explain how people behave in specific situations (Cropanzano and Mitchell 2005). In this research it is used to shed light on two constructs: the dependency of the foreign subsidiary management on its expatriates, and the subsidiary relationship in regard to its headquarters connection. This relationship is tested with the level of headquarter trouble.

Social Exchange Theory is not without its critics. This theory is a rather old one, established in the 1970s providing room only for a limited creative design (Miller 2005). Nowadays, knowing that there are many different groups in a single subsidiary, it is a difficult task to successfully apply this theory in the way it has been done previously. Also, there is heavy criticism about the often used five-point Lickert scale approach for being outdated.

However, Social Exchange Theory has been applied in combination to other constructs and has been applied at different occasions. The theory has been applied to constructs such as justice criteria in regard to the confidence level of firms (Colquitt and Rodel 2011). Kono (2011) applied it to leadership issues. Cameron and Webster (2011) shed light on effects of communication issues. Others, like Byrne et al. (2011) investigate to which extent trust influences the firm's performance. Khazanchi and Masterson (2011) also take this theory as their fundamental basis to research about justice and trust in firms.

3.3 Development of Hypotheses

The corner stone of the hypotheses developed in this third chapter are subsidiary dependency on expatriates and the relationship between the subsidiary and the headquarters.

The hypotheses are; therefore, composed of two different lines of argumentation. First argumentation covers subsidiaries and their dependency on their expatriates. In case a subsidiary depends heavily on their expatriates, it is assumed that in these cases, expatriates would not have suddenly left at the aftermath of the disaster. This is based on arguments of the Social Exchange Theory. A high dependency on expatriates means there is a need for social exchange between the subsidiary and the headquarters. That is, highly valued expatriates should remain in the aftermath of the disaster. Stated differently, expatriates would not departure these subsidiaries because the trust into them is rather high, which also has a positive influence on performance (Byrne et al. 2011; Khazanchi and Masterson 2011).

The second research stream covers the headquarter trouble. Each subsidiary has, to some extent, trouble or argumentation issues with the headquarters. This item is somehow connected to the independence of a subsidiary and their strategic orientation. Some foreign subsidiaries are quite independent against their headquarter politics, others rely heavily on and strongly work in accordance to the headquarters. There are many cases where the headquarter prefers a global strategy but the Japan based subsidiary feels that the Japanese market is different and that an individual strategy is necessary.

In case a subsidiary has higher trouble with the headquarters, it can be assumed that more expatriates would departure in the aftermath of the disaster. The reason for this is straightforward. Expatriates are sent for the headquarters and at the subsidiary level often seen as "ambassadors." If the subsidiary management has rather high trouble solving issues with the headquarters, it is reasonable to assume that rather more expatriates would departure the subsidiary in a case of a disaster. Out of the following, two hypotheses are offered:

H 1.1
The higher the dependency on expatriates, the fewer left the subsidiary at the time of the disaster.

H 2.2
The higher the HQ trouble, the more left the subsidiary at the time of the disaster.

Next, I investigated the difficulties for the subsidiary to keep expatriates in Japan. Logically, if the dependency on expatriates is rather high, it should be more difficult for the subsidiary to hold them in Japan. This is supported through the fact that subsidiaries who rely heavily on expatriates might face bigger problems when they departure. Stated differently, if dependency is high, not many of the expatriates should departure at the disaster. However, if they want to departure even while having a high dependency, it should be more difficult for the subsidiary management to hold them back.

Arguments for headquarter trouble as the dependent variable has similar origins. If the headquarter trouble is high, it should be more difficult to hold expatriates back to the subsidiary. There is reason to believe that a subsidiary management treats their expatriates as headquarter-sent managers. It makes sense to think that these expatriates who are treated as "outsiders" should not have the same hold at a Japan based subsidiary. They should prefer to departure home to the headquarters to Germany much faster than others who are more embedded into the subsidiary. Also, in the case that the connection between expatriates and headquarters is rather strong, there might be a communication problem between the subsidiary management and the headquarters. Out of these arguments, two hypotheses are developed.

H 2.1

The higher the dependency on expatriates, the more difficult to hold them back.

H 2.2

The higher the HQ trouble, the more difficult to hold them back.

Expatriates have important issues to deal with and often perform central roles at any given foreign based subsidiary. Many headquarters still today rely on their expatriates for various reasons. For example, one reason would be to get information from their subsidiaries, or to transfer information to their subsidiaries. Other functions are composed of controlling the subsidiary, or just to "show face" at the subsidiary abroad (Bruning et al 2011). That means at some subsidiaries, an expatriate's main activity is to demonstrate that the subsidiary is not a local firm, but an international player. In this sense, it is important to investigate to what extent subsidiaries faced problems because of their expatriate's departure. The next pair of hypotheses; therefore, will focus on problems subsidiaries faced because of a sudden retreat of their expatriates.

That suggests that a subsidiary with a rather high dependency on their expatriates would face more problems when expatriates departure it. Otherwise, when the dependency on expatriates is low, the impact of leaving expatriates should be rather low for the subsidiary.

Let us examine the hypotheses regarding headquarter trouble in another way. If the headquarter trouble is high, the subsidiary should face less problems in case expatriates departure it. To exaggerate, a subsidiary with high levels of headquarter trouble should be happy to see the expatriates leaving the subsidiary. Out of this argument, two hypotheses are derived.

H 3.1

The higher the dependency to Expatriates, the higher the problems of a sudden departure.

H 3.2
The higher the Headquarter trouble, the lower the problems of a sudden Expatriates departure.

The fourth pair of hypotheses are not directly connected to the aftermath of the disaster. To what extent do subsidiaries face problems attracting expatriates nowadays? As this study was conducted on March 2012, the opinions have a one year time lag. In regard to the Social Exchange Theory, subsidiaries whose expatriates suddenly returned back to Germany should have difficulty getting them back. Highly dependent firms may even face bigger problems in having expatriates come back. This is based on the fact that highly dependent subsidiaries lost their expatriates—despite the fact that they were highly dependent and did not want them to departure. A year after the disaster, it should be; therefore, more difficult to get expatriates back again.

The same relationship should count for the item headquarters trouble and the relationship to the expatriates. Where headquarter trouble is high, subsidiaries should face even more difficulty bringing expatriates back to Japan. The reason for this is that the headquarters is not supportive to send expatriates (back) to Japan. Also, because of a bad relationship between the headquarters and the subsidiaries, expatriates probably would not like to "jump into a venture" to become sandwiched between two different groups interests. Two hypotheses are developed:

H 4.1
The higher the dependency on expatriates, the higher the difficulties still presently remain to have them come back.

H 4.2
The higher the headquarter trouble, the more difficult to have expatriates return back to Japan (Fig. 3.1).

3.4 Methods

The methods section will start with explanation of the equilibrium models and then of the constructs. Then, the independent as well as the dependent variables are explained.

Equilibrium Model 1

Subsidiary dependency
= β0 + βiAge+βiSize(lg)+ βiReturningHome+βiRetainatSubsidiary+ βi Sudden DepartureProblem+βiTodayDrawExpatriates+ε

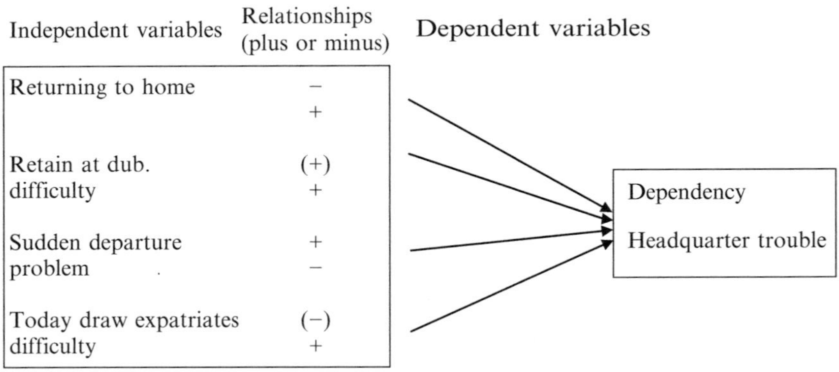

Fig. 3.1 Expatriate items on dependency and headquarter trouble

Equilibrium Model 2

Headquarter trouble
= β0 + βiAge+βiSize(lg)+ βiReturningHome+βiRetainatSubsidiary+ βi Sudden DepartureProblem+βiTodayDrawExpatriates+ε

Model 1 is about subsidiary dependency on expatriate and Model 2 is about subsidiary trouble to the headquarters. Both models have moderating variables, such as the year and the size of the Japanese subsidiary.

Independent variables consist of the item to which degree expatriates suddenly left the subsidiary (βiLeavingHome). The second item is the degree to which subsidiaries tried to hold expatriates back (βiHoldBack). The third item is the degree to which a sudden departure the subsidiary put into problems (βi Sudden Departure Probl.). The last item is about the degree to which subsidiaries presently face problems having expatriates come back to Japan (βiToday Back Japan). Two dependent variables are employed. The first variable is subsidiary dependency on expatriates. I ask how strongly subsidiaries depend on their expatriates. The second dependent variable is the degree of headquarter trouble a Japan based German subsidiary faces.

3.5 Results

This chapter deals with the results stemming from multivariate regression analysis in regard to German firm's impact of the Great East Japan Earthquake.

First, H 1.1 stated that the dependency on expatriate should influence whether or not these fellows would departure a subsidiary. The result shows to 1 % significance, that firms with a high dependency on expatriates experienced fewer departures from the subsidiary.

H 1.2 stated again a negative relationship. Subsidiaries with a higher headquarters trouble should face more expatriates going out of the office. While the coefficient is negative, as was assumed, the hypothesis failed to reach a level of significance.

H 2.1 dealt with the relationship of expatriate dependency and whether this was related to difficulties of keeping expatriates in the office. It can be seen that this is the case. The hypothesis is positive and reached a 1 % level of significance. This might be explained by the act that subsidiaries who depend on their (leaving) expatriates have more difficulties than other subsidiaries. H 2.2 deals about difficulties to held expatriates back and the level of headquarter trouble, a subsidiary faces. To 5 % significant is the relationship that subsidiaries with a higher headquarter trouble had more problems to hold them back. This result is also understandable, because expatriates are often seen as representatives of the headquarters.

H 3.1 deals with dependency of expatriates and to what degree the subsidiary ran into problems because of their sudden departure. In the descriptive analysis it was made clear that the impact of problems was rather low. However, firms with a high expatriate dependency also faced problems in cases where they left. This hypothesis was supported to 5 %. Also H 3.2, in regard to problems with expatriate departures and headquarter trouble, an astonishing negative relationship is visible. To a 5 % significance level the departure of expatriates at subsidiaries with a high headquarter trouble did not, in fact, cause problems for that subsidiary. Rather, those subsidiaries were relieved to see the expatriates go.

The last part of the hypothesis was about whether expatriates come back to Japan nowadays. H 4.1 (dealing with expatriate dependency) shows that it is significant to 5 % that firms with a higher dependency on expatriates have higher difficulties getting them back to Japan. H 4.2 (regarding headquarter trouble and the current expatriate situation) did not turn out to be statistically significant (Tables 3.1 and 3.2).

Table 3.1 Multiple regression with reliance to subsidiary as dependent variable

	Coefficient	SE	Significance
Age	0.00	0.01	0.906
Size (lg)	–0.04	0.25	0.88
Returning home	–0.60	0.20	***
Retain at subsidiary difficulties	0.60	0.21	***
Sudden departure problem	0.35	0.16	**
Today draw expatriates	0.27	0.13	**
_cons	1.16	0.42	***

43 Observations, $R^2=0.47$
*=10 %; **=5 %; ***=1 %

Table 3.2 Multiple regression with headquarter trouble as dependent variable

	Coefficient	SE	Significance
Returning home	–0.11	0.19	
Retain at subsidiary difficulties	0.53	0.20	**
Sudden departure problem	–0.44	0.18	**
Today draw expatriates	0.05	0.14	
_cons	1.74	0.42	***

52 Observations, $R^2=0.24$
*=10 %; **=5 %; ***=1 %

3.6 Discussion

This research is about two constructs. The first construct deals with the expatriate dependency of the Japan based German subsidiaries and on their impact. The second area of research is about the correlation of several expatriate items to subsidiary trouble with the headquarters.

The research shows that more than half of the subsidiaries claimed to have cases where expatriates suddenly left. The main reason why expatriates left the subsidiary was mainly the fear of radiation because of the Fukushima power plant accident. A second reason was the aftershocks that appeared frequently in the whole Kanto area. However, not all of the expatriates really went home to Germany. The interviews shed more light on this issue. Even if the question placed emphasis on whether expatriates left Japan or not, some of the expatriates went to places such as Hong Kong or Singapore instead. That is also a departure; however, the quality is arguably different. An expatriate who moved to Hong Kong is more likely to go back to the Japan office than an expatriate who was sent home to Germany. Some of these Germany-sent expatriates might not want to get expatriated anymore; some others might want to go to other parts of the world.

There is a statistically significant relationship between the sudden departure of an expatriate and their dependency. That was suggested in the hypotheses because, in subsidiaries with a higher dependency, expatriates know that they are needed. Their return back to Germany would put the subsidiary in trouble. It is arguable that expatriates in high dependency subsidiaries did not want to departure because they felt important at their positions. The second item addressed difficulties of keeping them. Also this item is statistically supported with a positive relationship. That means, firms who depend more on expatriates had higher difficulties to hold them back. If a firm does not depend on their expatriates, they should not have difficulties related to their departure in the time of a disaster. The third item looked at whether or not subsidiaries actually faced problems because of a sudden departure of expatriates. The positive relationship is clearly supported in interviews. Some firms with a high dependency faced big problems because of their expatriate departure. There is evidence that these firm's expatriates had higher positions at the subsidiary and there is some evidence that these firms also relied on their headquarter connection mainly through their expatriates. As they left, the subsidiary was unsure what to do. Even the fourth hypothesis was supported. It is visible that firms who rely on expatriates do not receive them in sufficient amount to Japan. There is some evidence suggesting that expatriates from German headquarters themselves prefer to go to other countries than Japan. One measure of performance is the increase of turnover at a given subsidiary, expatriates prefer to go to countries such as China or India, where there is an increase automatically. That means, China expatriated people would get preferable treatment in terms of promotions when going back to the headquarters. In contrast, Japan-sent expatriates would be rather beneficial to become the connection to the Japanese activities. As German business takes more hold in Asian countries other than Japan, the incentive to go to a country such as Japan is then rather low.

The reason why the impact of a departure was so low for the subsidiary is because many of the expatriates do not stay long enough at the Japan office. From the perspective of the locals, expatriates are seen as representatives of the headquarters.

Two significant relationships were found for headquarter trouble. First, subsidiaries with more headquarter trouble obviously faced more difficulties when trying to keep expatriates from returning to Germany. This was supported in many of the interviews, where some of the managers clearly pointed out that the difficult relationship with the headquarters could have been a reason that it was hard to keep expatriates in the aftermath of the disaster. As expatriates are seen much more as representatives of the headquarters than representing the local office, it was also no surprise that subsidiaries were actually happy when expatriates left of the office.

3.7 Conclusion

This research deals with the impact of the Great East Japan Earthquake on German firms. The descriptive results show that more than half of the subsidiaries experienced an impact due to the sudden departure of expatriates. It is furthermore evident that many firms faced difficulties in getting those departed expatriates to return back to the Japan-located subsidiary. Here, more than half of the firms reported difficulties in this area.

However, this sudden departure did not turn out to be a big problem for most of the firms. This means, subsidiaries did not face any serious problems when expatriates suddenly left. Subsidiaries reported, on average, a good relationship with their headquarters.

In the following multivariate regression analysis, I investigated to which extent subsidiaries' dependency correlated to expatriate departures and how the subsidiary-headquarter relationship correlated in terms of expatriate departures from the subsidiary. The results for expatriate dependency show that firms with higher dependency had fewer expatriates leaving the office. Also, the higher the dependency, the more difficult was it for the firms to bring expatriates back to Japan. A third understandable result is that the higher the dependency was, the more problems the subsidiary faced due to a sudden lack of expatriates. Also, subsidiaries with a higher dependency on expatriates report that it is still nowadays difficult to get expatriates back to Japan.

There are two results for the expatriate items in regard to the headquarter trouble. The higher the headquarter trouble was, the more difficult was it to hold expatriates back to the subsidiary. Also, firms with a higher rating of headquarter trouble reported lower problems with a sudden departure of expatriates.

This work is not without its limitations. The item headquarter trouble was measure with just one item. Also, top managers at the subsidiary might be not objective enough to fully count on their opinion. It can also be argued that not all expatriates went back home straight away. Some of them were sent to Singapore or Hong Kong. The registration office in Hong Kong e.g. reported a sharp increase of new registrations by non-Japanese who applied from Japan for a permanent Hong Kong visa. Two weeks after the disaster alone, there were 300 new applications reported of visa requests. That was the same amount than the whole 3 month before. The office; therefore, speeded up the procedure from 4 to 6 weeks for visa applications to only two days (Maechi 2011, p. 9). Out of 270 cases, it is reported that 158 were issued for a permanent work visa and 62 for families. It is also reported that people who moved from Japan to Hong Kong were in high positions. Out of these applications, more than 80 % earned more than 10,000 Euro monthly, mainly employed in the banking business.

In spite of its limitations, this work sheds some important light on expatriate departures in the aftermath of a disaster. It could be shown that the subsidiary dependency as well as the subsidiary headquarter relationship are important factors in forecasting at which of the firms expatriates left.

Appendix

Table 3.3 Means

	Mean	SE
Age	32.21	4.60
Size (lg)	1.76	0.13
Returning home	2.95	0.28
Retain at subsidiary difficulties	3.07	0.27
Sudden departure problem	1.81	0.18
Today draw expatriates	1.98	0.20
Reliance at sub.	2.33	0.19

43 Observations

Table 3.4 Correlations

Correlation	Reliance at sub.	Returning home	Retain at subsidiary difficulties	Sudden departure problem	Today draw expatriates	Headquarter trouble
Reliance at sub.	1					
Returning home	0.1142	1				
Retain at subsidiary difficulties	0.271	0.8443	1			
Sudden departure problem	0.5447	0.4819	0.5052	1		
Today draw expatriates	0.3614	0.2951	0.2885	0.3248	1	
Headquarter trouble	−0.0971	0.2746	0.3402	−0.1039	0.0361	1

83 Observations

References

Bebenroth R (2012) Umfrageergebnisse: Auswirkungen der Dreifachkatastrophe auf deutsche Unternehmen in Japan. Controller Mag (September)5:76–81

Blau PM (1964) Exchange and power in social life. Wiley, New York

Bruning NS, Bebenroth R, Pascha W (2011) Valuing Japan-based German expatriate and local manager's functions: do subsidiary age and managerial perspectives matter? Int J Human Res Manag 22(4):778–806

Byrne Z, Pitts V, Chiaburu D, Steiner Z (2011) Managerial trustworthiness and social exchange with the organization. J Manag Psychol 26(2):108–122

Cameron A-F, Webster J (2011) Relational outcomes of multicommunicating: integrating incivility and social exchange perspectives. Organ Sci 22(3):754–771

Colquitt JA, Rodel JB (2011) Justice, trust, and trustworthiness: a longitudinal analysis integrating three theoretical perspectives. Acad Manag J 54(6):1183–1206

Cropanzano R, Mitchell MS (2005) Social exchange theory: an interdisciplinary review. J Manag 31:873–900

Khazanchi S, Masterson SS (2011) Who and what is fair matters: a multi-foci social exchange model of creativity. J Organ Behav 32:86–106

Kono Y (2011)Ridashippu ron ni okeru soryo sakuyo apurochi no tenkai (Joint approach and evaluation of theory on leadership). Kanzai Daigaku Ronshu 56(3):41–53

Maechi M (2011) Gaishikei, Hong Kong he taihi (Foreign affiliated firms move to Hong Kong). Asahi Shinbun. 6 March, p 9

Miller K (2005) Communication theories. McGraw Hill, New York

Chapter 4
Closing Chapter

Abstract This empirical study about the impact of the Great East Japan Earthquake from March 11th 2011 on German firms is questionnaire-based with a response rate of around 35 % out of 244 firms, which were all located in the Kanto area. This study is complemented by information from top managers who were interviewed by the author at 14 different Kanto based German firms.

It reveals that the disaster had a big impact on German firms. Some of them changed their location, others temporarily closed. Even others with no motion of change externally faced internal problems. Half of all the firms confessed that their expatriates suddenly left their office. Even if the trouble because of their departure was not reported as a very heavy one, firms faced problems to bringing their expatriates back.

One year after the disaster, German firms seem to have recovered and all of the participants declared they now "operate as normal" in their subsidiary. Negatively affected were at least a quarter of the German firms; however another quarter of the respondents surprisingly saw a positive impact on their business caused by the disaster. Finally, half of all the respondents did not feel any kind of impact on their business because of the disaster. Some firms mentioned a positive and a negative impact at the same time, which balanced out. German firms did not close their Japan offices. There is, therefore, no fear that this disaster led to a relocation of German firms to other Asian countries. One could argue that only the "leftovers" were interviewed or responded to the questionnaires, all others might have already left. However, according to the German Chamber of Commerce at Tokyo, their members have not decreased beyond the normal fluctuation. The German subsidiaries enjoy, on average, a relatively high autonomy, which might have been established by a long standing presence in Japan.

Participants who reported that their expatriates suddenly left their subsidiaries (50 % of all participants) were further investigated during the follow up interviews. It became evident that the decision was not seldom influenced by factors other than

R. Bebenroth, *The Great East Japan Earthquake and Its Impact on German Firms: An Empirical Study*, SpringerBriefs in Business, DOI 10.1007/978-4-431-54451-7_4,

the expatriate him or herself. In some cases the families asked them to come back to Germany. In other cases, the German headquarters called them back to Germany. Also, the Ministry of Foreign Affairs was, in some cases, actively involved in getting expatriates back to Germany. In this booklet; therefore, the reason was not asked. Instead, top managers were asked about what their expatriates' departure caused at the Japan based subsidiaries. Besides the fact that many expatriates left the subsidiary, it is also evident that subsidiaries clearly had difficulties in getting them back. Short-term visitors even more frequently went back to Germany or, in many other cases, did not come again to Japan for the next month. Also a quarter of locally hired non-Japanese have left the subsidiaries at the aftermath of the disaster.

However, in all three investigated groups, the offices did not face serious problems because of their absence. This counts for expatriates and also for short-term visitors and even for their locally hired non-Japanese. The situation of non-Japanese employees today can be stated as normalized. Subsidiaries are also more optimistic nowadays to receive non- to relocate out of Tokyo (Kanto) area. It became evident that highly autonomous subsidiaries changed their status significantly more often.

The second chapter investigated how subsidiary autonomy and departure of expatriates impacted Japan-based foreign subsidiaries to either temporarily close down their offices in Kanto area or at the subsidiary was, the more often were temporary closings or relocation undertaken. Also, the bigger firms changed more often, as suggested by the hypothesis. However, it was the younger, not the older firms, which changed their status more often. The follow up interviews with top managers at 14 different Kanto based German firms made evident, that nowadays, a year after the disaster, a move out of Kansai area or even a temporary closing is seen as a costly mistake. Many of the managers report that they moved too fast and too early and should have waited in the aftermath of the disaster. According to transaction cost theory, which argues that subsidiaries with a higher level of autonomy have a better performance, I show that this reasoning does not work in a time of a disaster. At this specific time a rather lower level of autonomy and more communication to and with the headquarters office would be recommended.

The third chapter discusses expatriate dependency by subsidiaries and relationships between headquarters and subsidiaries. As stated above, for most of the subsidiaries a sudden departure of their expatriates did obviously not hurt them too much. However, it is evident that when firms had a higher dependency on their expatriates, fewer left the office. Also, higher expatriate-dependent subsidiaries faced more difficulties getting their expatriates back to their Japanese offices. Also, the more dependent subsidiaries faced more problems when expatriates.

Finally, nowadays, subsidiaries with a higher expatriate dependency report that it is still difficult to attract expatriates from their headquarters. There might be reasons apart from the disaster impact. Some interviewees pointed out that one of the important performance measures to get promoted at the headquarters is an increase in turnover. When expatriates would go to Japan, it can be considered to be a good performance already just to hold the previous year's result. At other countries, such as China or India, it would be quite easy to satisfy the headquarters' performance hurdle in terms of a turnover increase. Also, in terms of future promotions of expatriated managers, there is good reason to assume it is more attractive to get employed at a subsidiary in an emerging market rather than in an already developed and highly competitive country such as Japan.

Subsidiaries reported a relatively good relationship with their headquarters. The results at the third chapter show that when the subsidiary—headquarter relationship was in trouble, it was more difficult for the subsidiary management to keep expatriates in Japan. Also, when subsidiaries faced more trouble with their headquarters, problems decreased as expatriates suddenly left their Japan based offices. That result is supported by the interviews. Some of the subsidiaries with a difficult headquarter relationship; therefore, seemed relieved when expatriates left their offices to go home to the headquarters.

About the Author

Ralf Bebenroth was born and raised in Kassel, Germany. After graduating from compulsory high school he completed a 3 year internship (Ausbildung) as a locksmith. After that, he returned to school with aspirations for business school and began preparations for University. Ralf Bebenroth enrolled in the program of Business Administration at University of Kassel, and received his PhD in March 2001.

From April 2001 to March 2003 he was awarded a post doctoral fellow ship by DAAD, Monkasho at Tokyo Institute of Technology and conducted research about challenges in the integration period at Mergers and Acquisitions. From July 2003 to May 2005 he was again awarded a post doctoral fellowship, this time by Humboldt-foundation. Research was carried out at Osaka University of Economics about Corporate Governance issues. From 2005 Ralf Bebenroth began as Associate Professor at Kobe University in the Institute for Economics and Business Administration, teaching all classes at the faculty in Japanese language for the next 5 years. From 2009 he moved on to a tenured position as Associate Professor and from September 2012 was appointed to be Full professor. Although his research interests encompass a range of topics relating to the flow of human and financial resources into Japan and Japanese firms, his primary interest still remains cross-border issues of Mergers and Acquisitions.

R. Bebenroth, *The Great East Japan Earthquake and Its Impact on German Firms: An Empirical Study*, SpringerBriefs in Business, DOI 10.1007/978-4-431-54451-7,

Index

R. Bebenroth, *The Great East Japan Earthquake and Its Impact on German Firms: An Empirical Study*, SpringerBriefs in Business, DOI 10.1007/978-4-431-54451-7,

MIX
Papier aus verantwortungsvollen Quellen
Paper from responsible sources
FSC® C105338

If you have any concerns about our products,
you can contact us on
ProductSafety@springernature.com

In case Publisher is established outside the EU,
the EU authorized representative is:
Springer Nature Customer Service Center GmbH
Europaplatz 3, 69115 Heidelberg, Germany

Printed by Libri Plureos GmbH
in Hamburg, Germany